The INSIDE GUIDE

BIOLOGY BASICS

Plant Parts and Photosynthesis

By Leigh McClure

Cavendish Square

Published in 2025 by Cavendish Square Publishing, LLC
2544 Clinton Street Buffalo, NY 14224

First Edition

Website: cavendishsq.com

Cataloging-in-Publication Data

Names: McClure, Leigh.
Title: Plant parts and photosynthesis / Leigh McClure.
Description: Buffalo, NY : Cavendish Square Publishing, 2025. | Series: The inside guide: biology basics | Includes glossary and index.
Identifiers: ISBN 9781502673206 (pbk.) | ISBN 9781502673213 (library bound) | ISBN9781502673220 (ebook)
Subjects: LCSH: Plant anatomy–Juvenile literature. | Photosynthesis–Juvenile literature. | Plant life cycles–Juvenile literature.
Classification: LCC QK64.1 M28 2025 | DDC 581.4–dc23

Editor: Caitie McAneney
Copyeditor: Nicole Horning
Designer: Deanna Lepovich

The photographs in this book are used by permission and through the courtesy of: Cover fotohunter/Shutterstock.com; p. 4 Belikova Oksana/Shutterstock.com; pp. 6, 12, 27 BlueRingMedia/Shutterstock.com; p. 7 Joachim Stingl/Shutterstock.com; p. 8 K.Kargona/Shutterstock.com; p. 9 (main) Panga Media/Shutterstock.com; p. 9 (inset) Stephen Moehle/Shutterstock.com; p. 10 malshkoff/Shutterstock.com; p. 13 Matthew James Ferguson/Shutterstock.com; p. 14 Santiago Quintero Alzate/Shutterstock.com; p. 15 Mega Vectors/Shutterstock.com; p. 16 Trp.Ssb1979/Shutterstock.com; p. 18 (main) Anshuman Rath/Shutterstock.com; p. 18 (inset) Viktor Sergeevich/Shutterstock.com; p. 19 Nicholas J Klein/Shutterstock.com; p. 20 CrizzyStudio/Shutterstock.com; p. 21 Holly S Cannon/Shutterstock.com; p. 22 eyetravelphotos/Shutterstock.com; p. 24 Thomas B Lee/Shutterstock.com; p. 25 Johannes Kraak/Shutterstock.com; p. 26 PARALAXIS/Shutterstock.com; p. 28 (top) imageBROKER.com/Shutterstock.com; p. 28 (bottom) Bilanol/Shutterstock.com; p. 29 (top) Halfpoint/Shutterstock.com; p. 29 (bottom) Riccardo Zamboni/Shutterstock.com.

CPSIA compliance information: Batch #CWCSQ25: For further information contact Cavendish Square Publishing LLC at 1-877-980-4450.

Printed in the United States of America

CONTENTS

Giant sequoia trees are the largest plants on Earth. The largest of them all is the General Sherman Tree in California.

THE WILD WORLD OF PLANTS

Chapter One

What do giant sequoia trees have in common with the dandelions in your yard or the moss on a rock? They're all plants! Earth is home to around 435,000 different species, or kinds, of plants. Some are small enough to fit in your hand, while others are some of the largest organisms, or living things, on Earth.

Plants have adapted over hundreds of millions of years to become the living things we see today. Some live in cold climates, while others thrive in warm, wet climates. Others, such as cacti, can survive in areas of Earth that are much too dry for other organisms. Let's explore the wild world of plants!

Plant Cells

How can you tell if something is a plant or an animal? You can look at its physical features—its color, shape, and parts. Does it stay put in the earth? Does it make its own food? You can also look at its cells under a microscope.

> **Fast Fact**
>
> The General Sherman Tree is 52,508 cubic feet (1,487 cubic meters). That's the size of a 26-story building!

Plant and animal cells are different from one another.

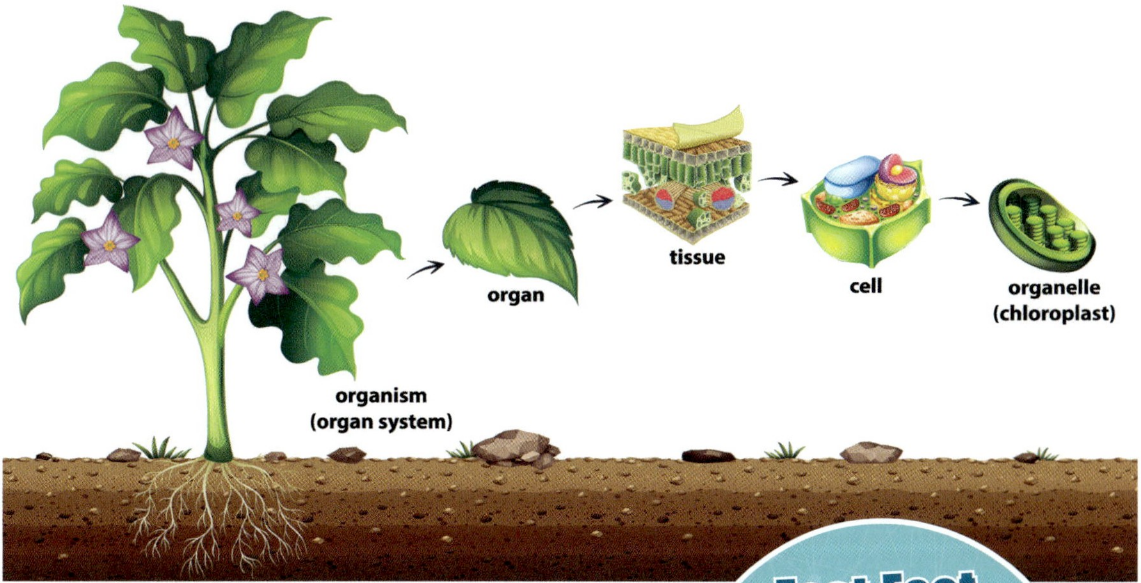

This cell diagram shows where chloroplasts are found within a plant's cell.

> **Fast Fact**
> Chloroplasts have a **pigment** called chlorophyll, which gives the cell and the plant a green color.

While animal cells may be different shapes, plant cells are rectangular. The cells have organelles, or structures, that carry out certain functions within the cell. The nucleus holds the cell's DNA, or deoxyribonucleic acid, which is like a blueprint for what the plant will look like and how it grows. While animals eat food, plants have special organelles called chloroplasts which help them make their own food using sunlight.

Vascular vs. Nonvascular Plants

Most plants you see are vascular. That means they have special tissues called xylem and phloem that send water and nutrients throughout the plant. Vascular plants take in water from their roots. These plants include trees, flowering plants, ferns, and more.

Is It Vascular?

Yes	No
coniferous trees	hornwort
deciduous trees	liverwort
flowering plants	moss
ferns	
grasses	
shrubs	

To tell if a plant is vascular, check for three essential parts: roots, stems, and leaves.

Fast Fact

Mosses are found in shady, moist places. They help the environment by absorbing water and giving nutrients to plants around them.

Nonvascular plants are low to the ground. They don't have roots, stems, or leaves. They include liverwort, hornwort, and moss. Most don't have tissues that transport water and food throughout their body. Instead, they absorb water through membranes.

This tree is vascular, while the moss covering it is nonvascular.

UNDERWATER PLANTS

Plants that live underwater are called aquatic plants. Some live in fresh water, while others live in salt water. They are different from algae, which is actually a plantlike organism called a protist. Some aquatic plants live completely underwater, while others float on the water. Some have root systems in the ground but leaves on the water's surface.

Aquatic plants are necessary for their ecosystem. They give other organisms in their underwater ecosystem the food, oxygen, and shelter they need. For example, many creatures hide in seagrass. It also helps keep the sea floor from **eroding**. Aquatic plants can even provide a safe place for fish to lay their eggs.

One of the smallest plants on Earth is duckweed. This plant looks like tiny dots on the surface of water. Amphibians use it for protection, while ducks use it for food.

Plant Diversity

There is a great variety of plants on Earth. Some produce flowers and fruit. Some float on the water's surface. Some lose their leaves in the winter, while others stay green all year. Their adaptations ensure that they can survive in different habitats and fit into their **ecosystem**.

The great differences in Earth's organisms is called biodiversity. It's very important to the well-being of ecosystems around the world. Whether they are large or small, plants provide shelter, protection, food, and materials for other organisms around them. That's why the biodiversity of plants must be protected in today's world.

While the General Sherman Tree is the largest tree by volume, the tallest tree on Earth—a coastal redwood called Hyperion—is 380 feet (115.8 m) tall!

Forests are habitats with many trees and plants. The Amazon Rainforest alone is home to 80,000 species of plants!

Root hairs are part of a plant's root system.

PLANT PARTS

Chapter Two

Think about the plants you see daily. Maybe you have a tree by your school or a shrub in your yard. You might have a potted orchid or a tomato plant in your garden. It's easy to find the differences between them, but can you get to their common roots?

The world is full of plants of many different sizes, shapes, and abilities. Some grow flowers and fruits, while others don't. However, most plants have a few basic parts in common. They have roots that go into the ground, a stem for structure, and leaves for photosynthesis.

Down to the Roots

Unlike animals, plants don't move from place to place. Instead, they are anchored into the ground by their roots. These underground plant parts keep plants from blowing or washing away. Tree roots dig deep into soil to hold the tree in place. Meanwhile, some flowering plants like black-eyed Susans and dandelions have very shallow root systems and are easy to pull out of the ground.

Fast Fact

Roots have a layer called the cortex that can store extra food for the plant.

Roots also gather water and nutrients from the soil. When it rains, roots suck up the water and send it up to the rest of the plant. The plant uses that water to perform its functions and stay alive.

Roots benefit more than just their plant. They help prevent flooding and erosion in their environment, or natural world. Some roots are eaten as vegetables, such as carrots, beets, and turnips.

Stellar Stems

How does the water get to the rest of the plant? This is a job for the stem. Water and nutrients from the soil are moved upward by xylem cells. The food made by leaves moves throughout the plant through phloem cells.

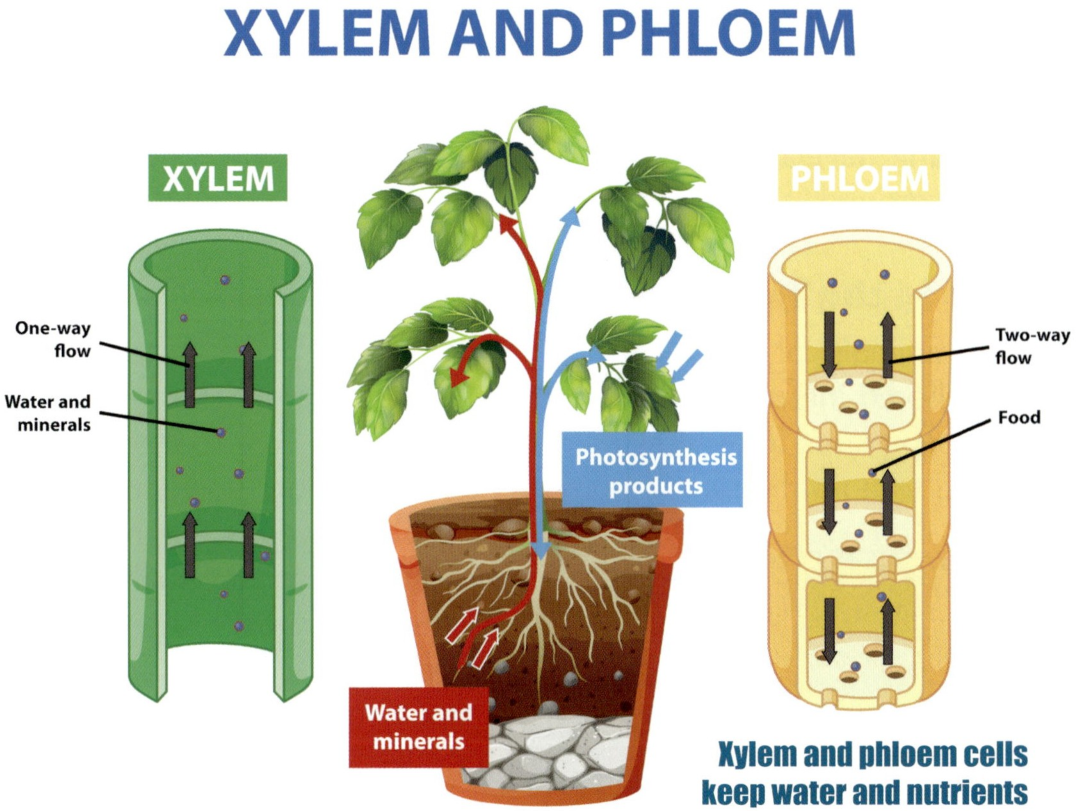

Xylem and phloem cells keep water and nutrients flowing within a plant.

Fast Fact

To deal with dry desert conditions, cacti store a great amount of water in their stems.

Stems also support the plant. Some smaller plants have soft, short stems. Tall plants, such as redwood trees, have woody stems called trunks. The largest are wide enough to drive a car through! Stems are a connecting place for the leaves, roots, and flowers. Water and food can be stored here too.

The saguaro cactus can grow taller than 45 feet (13.7 m). Its stem can hold more than 1,000 gallons (3,785.4 liters) of water!

FLOWERS

Flowering plants are called angiosperms. Some have small, dull flowers, while others have large, colorful flowers. A flower's bright colors attract **pollinators** that take and leave pollen as they go.

In an angiosperm, seeds grow inside a part called an ovary that's inside a flower. These plants also produce fruit. Some flowering plants produce juicy fruits such as mangoes and strawberries. Others produce hard, small, or **inedible** fruits. Many flowering plants are seasonal, meaning they grow and lose flowers at specific times of the year.

Earth is home to around 300,000 species of angiosperms.

A Look at Leaves

Some plants, such as fir trees, have small, needle-like leaves. Others have very wide or long leaves. Leaves come in many different shapes and color variations. The venation, or pattern of veins that transport water and nutrients, also differs from plant to plant.

Fast Fact

The raffia palm tree has the largest leaves in the world. They can be more than 65 feet (20 m) long!

Leaves are like little food factories for each plant. They are the places where photosynthesis happens. Leaves absorb carbon dioxide and sunlight, then convert the sun's energy into food for the plant through photosynthesis. The food is then sent throughout the plant. Small pores, or openings, in the leaves let oxygen out. This oxygen improves the air quality for animals and humans.

Each of these leaves features different shapes and venations, but they are all manufacturing food for their plant.

Zinnias are annual flowers. Their life cycle lasts one year.

THE LIFE CYCLE OF PLANTS

Chapter Three

As with all organisms, plants follow a life cycle. They are born, may **reproduce**, and then die. Some plants have short lives. For example, "annuals" are flowering plants that complete their life cycles in one year. Others have longer lives, such as large trees.

All plants start their lives as a seed or spore. They contain the chromosomes that have the genetic information necessary for a new plant to grow.

Plant Reproduction

Most vascular plants start their life cycles as seeds. Seeds contain two sets of chromosomes (one from a male cell and one from a female cell) and enough food stores to allow a baby plant to grow. Seeds are made through a process called sexual reproduction.

In a conifer (a tree with needle-like leaves), male cones make pollen. The pollen then makes its way by wind to female cones to make seeds.

In angiosperms, pollen that contains male cells and is

Fast Fact

The oldest living organism on Earth is a bristlecone pine tree in California named Methuselah. It's more than 4,700 years old!

Pollination

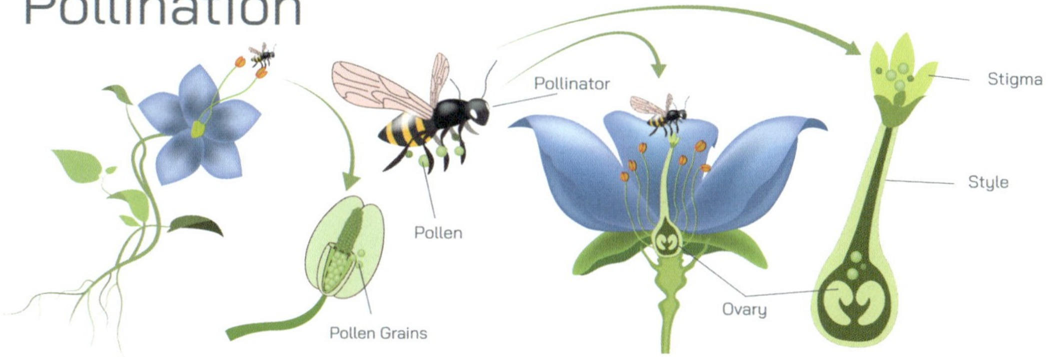

This diagram shows how pollination happens, forming new flowering plants.

produced in a flower's anther combines with female cells from a flower of the same plant or another plant. This process often happens with the help of pollinators, such as bees or butterflies. They land on a flower to drink a sweet liquid called nectar. In the process, the pollinator gets pollen on it, then brings the pollen to another flower. The pollen makes its way to the other flower's pistil, eventually reaching the ovary. There, it **fertilizes** eggs called ovules. A fruit starts to form, which has seeds within it. The seeds are often **dispersed** through animals that eat the fruit.

Potatoes are tubers that start to "sprout" after a while. The new plants are clones, or exact copies, of the parent plant. This is called vegetative propagation.

Fast Fact

Some plants don't need to join male and female cells together to form a new plant. They grow new plants from underground stem parts called tubers or rhizomes. That's a kind of asexual reproduction.

SUPER SPORES

Spores are the starting point for plants like mosses and ferns. Plants that reproduce this way have parts called sporangia that let out spores. These tiny spores typically have just one cell, but they can develop into a new plant. Once the sports are dispersed, they may land in rich soil. Then, they form tiny green **gametophytes** with female and male reproductive cells, called gametes. The cells come together to produce a sporophyte with two sets of chromosomes. The plant grows and then produces its own spores through asexual reproduction. The life cycle begins again.

Ferns and mosses are often plentiful in forest environments. Spores are dispersed by wind, rain, and animals, which are common in forests.

Growing Up and Out

If a seed has enough water and rich soil, it will break open. A small green plant will **emerge**. It pushes up through the soil to the sun as the roots creep down into the soil to anchor the plant.

Sprouts continue to grow upward, with their stems either becoming thicker to support themselves or climbing up other plants for support. Tree and shrub stems begin to harden, becoming woody. Eventually, plants grow their own cones, flowers, or buds, and start to reproduce.

A young tree is called a sapling.

Returning to Earth

In annual plants, life ends soon after reproduction. The plant drops its seeds and dies. In other plants, the top growth dies, but the roots remain alive and ready for the next growing season. Some other plants live hundreds or thousands of years. Eventually, though, a plant may reach the end of its natural lifespan, be cut down, or lack resources—sunlight, water, or space—to keep growing.

The plant will fall to the ground, where it will start to decompose, or break down. Its nutrients will return to the earth, making the soil rich for new seeds.

Fast Fact

An upright dead tree is called a snag. When it falls, it's a log.

Logs are often homes for wildlife, like this young badger.

Some plants live in places with a lot of sunshine. Others have adapted to areas with less sunshine.

FANTASTIC PHOTOSYNTHESIS

Chapter Four

Imagine you could make your own food out of sunlight. It would seem like a superpower! Plants are special because they all have the power to make the food they need to survive. This power is woven into their cells.

Photosynthesis happens within cell parts called chloroplasts. With the right amount of sunlight and water, a plant can thrive. When plants thrive, Earth thrives. Animals use the energy from plants to live too.

Using Sunlight

Plants need certain raw materials to make food. One is a substance found within chloroplasts called chlorophyll. It absorbs sunlight. Then, the plant needs a gas called carbon dioxide that's in the air. We breathe out carbon dioxide all the time. The roots of the plant suck up water and nutrients from the soil, which the plant also needs.

Fast Fact

Plants aren't the only organisms with chlorophyll. Algae and cyanobacteria also have chlorophyll and can produce food.

Chlorophyll makes leaves green. It occurs mostly in the leaves, but also in the stem.

These raw materials make up the perfect recipe for plants. The sun's energy changes carbon dioxide and water into a sugar called glucose. It also produces oxygen, which is let out through the plant's pores.

Follow the Food Chain

Plants are called producers because they produce their own food, which becomes a source of energy for animals, called consumers, in an ecosystem. Plants are the first part of a food chain.

Fast Fact
Elephants eat around 350 pounds (158.8 kilograms) of plant material per day!

Primary consumers are herbivores, or plant eaters, and omnivores, or animals that eat both plants and animals. Herbivores include small animals such as rabbits as well as larger animals such as elephants. After primary consumers are secondary consumers, which are meat eaters that eat primary consumers, such as the coyote that eats a rabbit. The energy from the sun is transferred from the plant to the rabbit to the coyote. Each step in the food chain is called a trophic level. The amount of energy from sunlight decreases with each trophic level.

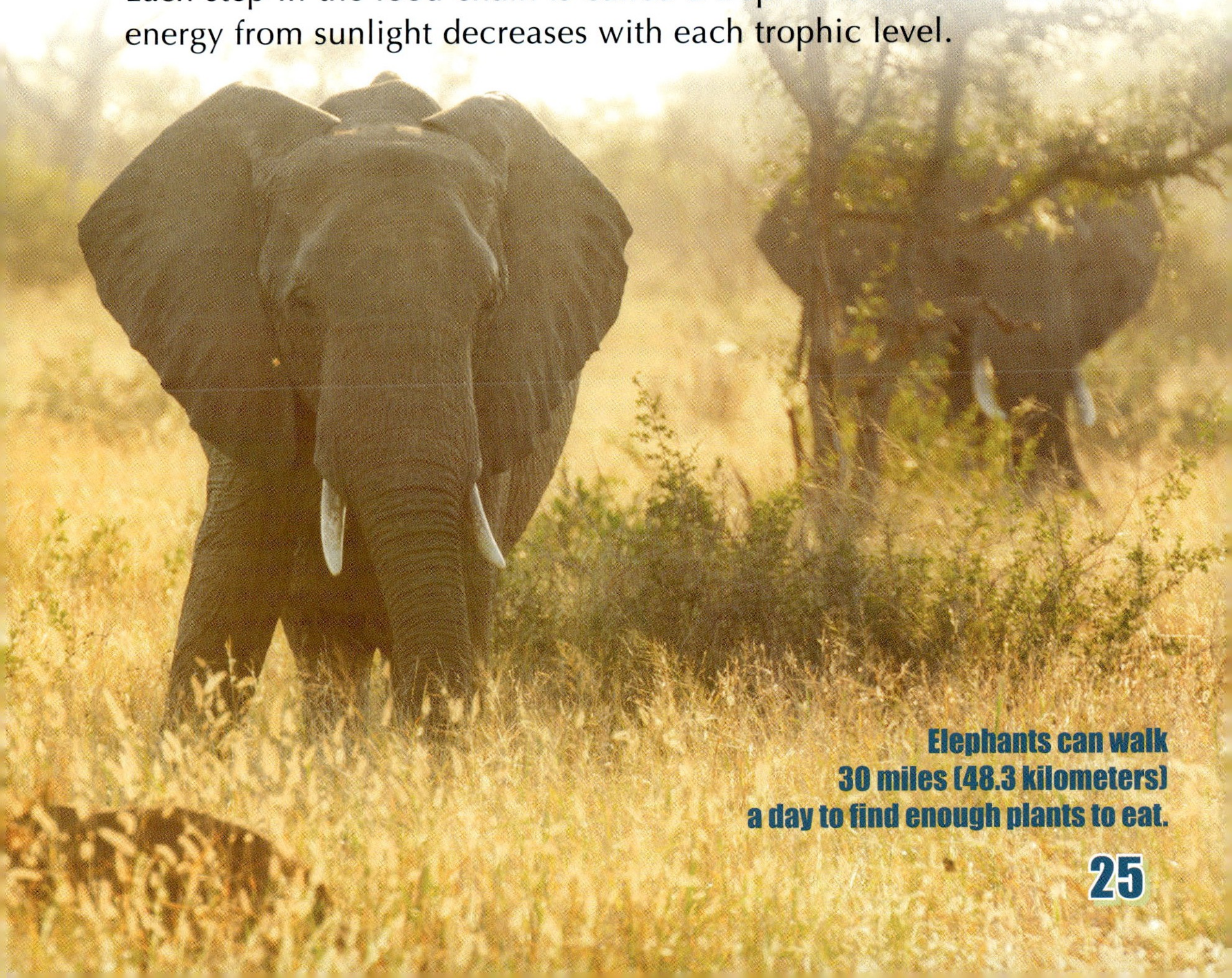

Elephants can walk 30 miles (48.3 kilometers) a day to find enough plants to eat.

Giving Back

Every living thing on Earth benefits from plants' ability to carry out photosynthesis. Each time you bite into a piece of fruit or a vegetable, the plant is giving you energy to carry out your daily tasks. Plants also provide the oxygen that all animals and humans need to survive. A world without plants would be unlivable.

However, humans have caused a great deal of harm to Earth's plants. They have cut down forests for lumber and other resources. They have polluted the air, water, and soil that plants need to survive. They have built homes, businesses, and farms on lands that once belonged to thick vegetation.

We need trees to survive **climate change** and its effects. They help prevent erosion, absorb excess carbon dioxide, and keep animals in their habitats. They give us food and materials, which we can use in **moderation**, thinking of the **sustainability** of our actions. Humans can give back—just as plants have given to us.

Around 17 percent of the Amazon Rainforest has been lost to deforestation and other human causes.

THE LUNGS OF THE EARTH

Rainforests are areas of thick vegetation in warm, wet climates. They're home to many species of animals, from macaws to tree frogs to jaguars. They're also home to many different kinds of trees. The tallest trees make a canopy over the others, creating the perfect conditions for animals that live underneath. The trees block out so much sun that the forest floor is dark.

Rainforests produce around 20 percent of the world's oxygen. So much oxygen is produced in the Amazon Rainforest that it is called the "lungs of the Earth." However, the Amazon Rainforest is in danger from deforestation, wildfires, and climate change.

Fast Fact

Rainforest trees called kapok trees can grow more than 200 feet (61 m) tall, blocking the sun from the trees below.

The tallest trees in the rainforest make up the emergent layer. That's followed by the canopy, the understory, and the forest floor.

THINK ABOUT IT!

1. Can you name examples of ways plants have adapted to their environments?

2. What effects might climate change and extreme weather have on plants?

3. Why are plants an
essential part of a
food chain?

4. How do flowering
plants and pollinators
help one another?

GLOSSARY

climate change: Change in Earth's weather caused by human activity.

disperse: To send out or scatter in all directions.

ecosystem: All the living things in an area.

emerge: To come out of a place.

erode: To wear away the earth's surface.

fertilize: To have reproductive cells come together to produce offspring.

gametophyte: In the life cycle of plants that reproduce using spores, the gamete-producing phase where one set of chromosomes is present.

inedible: Not safe to be eaten.

moderation: The avoidance of excess or extremes.

pigment: Anything that is used or serves to provide color.

pollinator: An organism that carries pollen to a plant.

reproduce: To make a new organism.

sustainability: A method of managing or using a resource so that the resource is never used up or forever damaged.

FIND OUT MORE

Books

Culliford, Amy. *Forest Animal Habitats*. St. Catherines, ON: Crabtree Publishing, 2025.

McKenzie, Precious. *The Micro World of Animal and Plant Cells*. Oxford, UK: Raintree Publishers, 2023.

Owen, Ruth. *Photosynthesis*. Minneapolis, MN: SilverTip Publishing, 2024.

Websites

Photosynthesis
www.ducksters.com/science/photosynthesis.php
Read more about how plants make their own food through photosynthesis.

Plant Problem | Hero Elementary
ny.pbslearningmedia.org/resource/parts-of-plants-media-gallery/hero-elementary/
Learn about plant parts with this fun and educational cartoon.

Redwood
kids.britannica.com/kids/article/redwood/390841
Explore the tallest plant in the world—the redwood.

Publisher's note to educators and parents: Our editors have carefully reviewed these websites to ensure that they are suitable for students. Many websites change frequently, however, and we cannot guarantee that a site's future contents will continue to meet our high standards of quality and educational value. Be advised that students should be closely supervised whenever they access the internet.

INDEX